AF619591

DIGITAL FORTRESS

PROTECTING YOUR DIGITAL IDENTITY

YOGESH UKEY

INDIA • SINGAPORE • MALAYSIA

ISBN 979-8-89026-829-7

HC: 979-8-89556-133-1

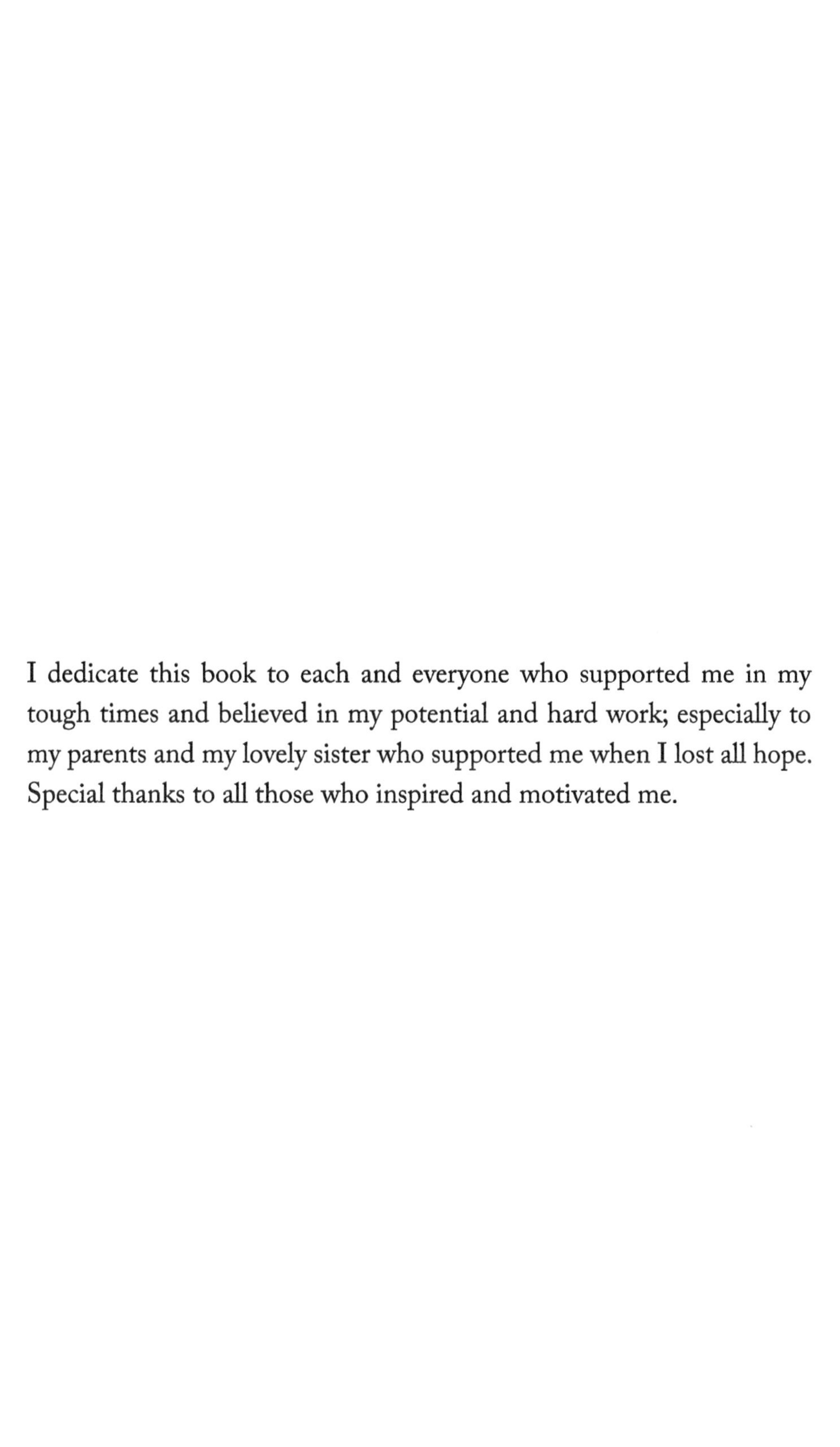

I dedicate this book to each and everyone who supported me in my tough times and believed in my potential and hard work; especially to my parents and my lovely sister who supported me when I lost all hope. Special thanks to all those who inspired and motivated me.

Contents

Preface

ATTENTION!

Don't worry, you can return all your files!

All your files like pictures, databases, documents and other important are encrypted with strongest encryption and unique key.

The only method of recovering files is to purchase decrypt tool and unique key for you.

This software will decrypt all your encrypted files.

What guarantees you have?

You can send one of your encrypted file from your PC and we decrypt it for free.

But we can decrypt only 1 file for free. File must not contain valuable information.

You can get and look video overview decrypt tool:

https://we.tl/t-72VNKmoPkb

Price of private key and decrypt software is $980.

Discount 50% available if you contact us first 72 hours, that's price for you is $490.

Please note that you'll never restore your data without payment.

Check your e-mail "Spam" or "Junk" folder if you don't get answer more than 6 hours.

To get this software you need write on our e-mail:

helpmanager@mail.ch

Reserve e-mail address to contact us:

restoremanager@airmail.cc

Your personal ID:

0263pErfnHcEhfdYJgbj69Em7z5YcG26owS6p4OdFxP5YTZG

One morning, my friend called and told me that when he opened his laptop, he received the above message as a text file on his desktop and in every folder. He wasn't able to open any of his files. All the extensions were converted to a different format. After doing some research on the Internet, he discovered that he had been the victim of a ransomware attack.

At the point, when his system had been infiltrated, he felt mistreated and helpless, unable to access his own information without paying a hefty payment.

As an ethical hacker, I know directly the damage that cyber security problems can cause. It was a reminder that caused me to recognize how fragile we are in this digital age. Because of this event, I chose to write this book to bring issues to your attention about network security and the basic demand for solid protection against digital risks. I want to share what I have learnt and assist other people keep away from the frustration and disappointment that my friend went through.

The world we live in today is getting increasingly networked and computerized, and thus, cyber security has become a core concern for people, companies, and nations. The rapid growth of innovation has created new faults and risks that must be addressed to preserve the safety and security of our own data, financial resources, and public safety. As we depend more on innovation for correspondence, monetary exchanges, and even medical services, the requirement for good digital security measures turns out to be more important. Cybercriminals are turning out to be more complex, and their attacks are also becoming more frequent and dangerous. The effects of a successful digital attack can be significant, including loss of delicate information, monetary loss, and harm to an organization's standing.

In this book, I will cover the main criteria of cyber safety, the threats and flaws that exist in the modern world, and the tactics and innovations utilized to prevent them. We will address issues, such as personal safety, organizational security, safeguarding data, and learning cyber security basics. We will likewise investigate the legal and moral difficulties around network safety, including security, data security, and digital warfare.

This book is expected for any person who wants to learn about cyber security, from students and IT experts to business pioneers and strategy creators. Especially, this book is for every father and mother whose children are always online. I hope that after reading this book, you will have a broader understanding of cyber security and the essential role it plays in modern society.

Important Note

This book is only for educational purposes. Hacking is a crime and should not be performed. You are responsible for your own acts. Neither I nor the book is responsible for any kind of acts you perform. This book is just a comprehensive guide to cyber security for general awareness only.

The Beginning: First Cybercrime Attack

In the early 1970s, a new and fascinating technology called the Internet was beginning to take shape. It held immense potential for connecting people, sharing information, and revolutionizing the way we interacted with the world. Little did anyone know that this digital world would soon become a playground for mischief and chaos.

In a small university town, nestled amidst the buzzing halls of Cornell University, lived a brilliant but mischievous graduate student named Robert Tappan Morris. Robert had always been fascinated by computers and possessed a deep understanding of their inner workings. He spent countless hours exploring the details of the emerging Internet, seeking to unravel its secrets.

One fateful day, in November 1988, Robert's curiosity got the better of him. He planned an experiment, a way to measure the size and strength of this rapidly expanding digital network. With a mischievous sparkle in his eye, he unleashed what would later be known as the 'Morris Worm' upon the unsuspecting world.

The Morris Worm was a devious creation, a self-replicating programme designed to spread from one computer to another, much like a biological virus. Robert had intended it to act as a harmless survey, exploring the interconnected web and reporting back with valuable information. But as fate would have it, a tiny coding error transformed his seemingly harmless creation into something far more sinister.

The Worm quickly gained momentum, exploiting vulnerabilities in the Unix operating system that powered many of the early Internet-connected computers. It exploited weak passwords, infiltrated systems, and replicated itself, launching multiple copies to propagate its presence.

As the Morris Worm continued its relentless march across the digital world, the consequences became apparent. The infected computers began to slow down, groaning under the weight of this unintended invasion. Critical services ground to a halt, and panic spread through the digital realm.

System administrators scrambled to contain the damage, but the Worm was relentless. It had become a digital wildfire, devouring unsuspecting machines and disrupting network communications. What was meant to be a harmless experiment had spiralled out of control, leaving chaos and confusion in its wake.

News of this extraordinary cyber-attack spread like wildfire, capturing the attention of governments, security experts, and computer enthusiasts worldwide. The Morris Worm had become the first documented cybercrime attack, an important moment in the history of digital security.

In the aftermath of the chaos, Robert Tappan Morris found himself facing the consequences of his actions. He was prosecuted under the newly established Computer Fraud and Abuse Act (CFAA), becoming the first person to be convicted under this ground-breaking legislation. The incident served as a sobering reminder of the need for strict cybersecurity measures and prompted a collective awakening to the vulnerabilities that existed within this brave new digital world.

The Morris Worm forever changed the atmosphere of cybersecurity, triggering a wave of research, policy development, and technological advancements. It highlighted the urgent need for stronger defences, heightened awareness, and collaborative efforts to protect against the ever-evolving threats lurking in cyberspace.

And so, the first documented cybercrime attack serves as a cautionary tale, a reminder that even the brightest minds must use their knowledge responsibly in the digital realm. It stands as a milestone in the ongoing battle to safeguard our interconnected world from the dark forces that seek to exploit its vulnerabilities.

Chapter 1

Introduction to Digital Security: Understanding the Risks

In today's interconnected world, when technology is a vital part of our daily lives, it is crucial to grasp the importance of cybersecurity. Cybersecurity refers to the procedures, measures, and strategies adopted to protect our digital systems, networks, and data from unauthorized access, theft, and damage. This brief guide seeks to illustrate the significance of cybersecurity in simple terms, especially for individuals without technical experience.

Points to Remember

The Digital Era

We are living in the digital era, where technology has altered the way we communicate, work, and interact. Our personal and professional lives are linked with digital platforms, ranging from smartphones and laptops to online banking and social networking. This digital dependence gives us ease but also exposes us to different cyber hazards.

The Growing Threat Environment

Cyber risks, such as hacking, identity theft, and virus attacks, are on the rise. Hackers and fraudsters constantly seek weaknesses in our digital systems to exploit for personal gain or harm. The consequences can be serious, including financial loss, privacy breaches, reputational damage, and even physical harm in some situations.

Personal Data Protection

We disclose a substantial amount of personal information online, including financial details, addresses, and even intimate elements of our lives. Cybersecurity guarantees that this sensitive data remains confidential and safeguarded. By employing rigorous security measures, such as strong passwords, encryption, and two-factor authentication, we can secure our personal information from getting into the wrong hands.

Financial Security

With the rise of e-commerce and online banking, our financial activities are increasingly handled electronically. Cybersecurity plays a critical role in securing these transactions and preventing unwanted access to our bank accounts, credit card details, and other financial assets. It helps safeguard the integrity and confidentiality of our financial information, decreasing the danger of financial fraud and identity theft.

Business Protection

In addition to personal security, cybersecurity is critical for organizations of all kinds. Companies keep huge volumes of sensitive data, including client information, trade secrets, and proprietary data. A successful cyber assault can disrupt operations, lead to financial losses, and damage a company's brand. Implementing robust cybersecurity is crucial to preserve these precious assets and keep the trust of customers and business partners.

National Security

Cybersecurity is not simply an issue for individuals and organizations; it is a subject of national security as well. Critical infrastructure, such as power grids, transportation systems, and government networks, are possible targets for cyber-attacks. Breaches in these systems can have significant implications, including disruption of key services and compromise of sensitive government information. Strong cybersecurity policies are necessary to defend the nation's key infrastructure and ensure its overall security.

Protecting Privacy

In today's digital age, privacy is a priceless commodity. Cybersecurity helps protect our privacy by shielding our personal information from illegal access. It ensures that our communications stay private and safe, shielding us from surveillance, cyberstalking, and intrusive data collecting. By emphasizing cybersecurity, we can control our digital lives and protect our right to privacy.

As technology continues to evolve, the necessity of cybersecurity should be emphasized. Individuals, corporations, and governments need to understand the risks posed by cyber threats and take proactive steps to defend themselves. By installing robust security measures, being watchful against potential attacks, and consistently educating ourselves about best practices of cybersecurity, we can build a safer digital environment for everyone. Remember, cybersecurity is not merely a technical concern; it is a shared obligation to secure our digital environment.

In the modern age, innovation has turned into a fundamental piece of our lives. From virtual entertainment to web-based banking, we depend on computers and mobiles for financial transactions, shopping, work, and engaging ourselves. Nonetheless, with the accommodation and advantages of innovation, we are prone to dangers to our data and security. In this part, we will see in detail the essentials of cybersecurity, the most widely recognized dangers, and how to protect ourselves from them.

Here are some questions we must know to understand better why we need cybersecurity in the modern digital age.

What Are Some Common Misconceptions About Digital Security?

There are several common misconceptions about digital security that can prevent individuals and organizations from effectively safeguarding their digital assets. Some of these misconceptions include:

1. **"Cyber security is only a concern for large organizations or governments."** In reality, cyber attackers do not discriminate based on company size or industry. Small businesses and individuals are just as likely to be targeted, and a lack of robust security measures can make them easier targets.
2. **"Investing in sophisticated security tools is enough to keep us safe."** While investing in high-end security tools and solutions is an essential part of keeping your business secure, it won't protect you from everything. Security tools and solutions are only fully effective if they are properly configured, monitored, maintained, and integrated with overall security operations.
3. **"Staying compliant with industry regulations is enough to keep us safe."** While staying compliant with industry data regulations is essential for doing business, establishing trust and avoiding legal consequences and regulations often provide only the bare minimum of security practices. Being compliant does not mean you are secure.
4. **"Cyber security is the responsibility of the IT department."** In reality, cyber security is the responsibility of every team member; from the founder who sets a security-minded tone to the teams that implement the policies to the new employee choosing a password.

By being aware of these common misconceptions and taking a proactive approach to digital security, individuals and organizations can better protect themselves from cyber threats.

What is Cybersecurity?

Cybersecurity or Digital security, otherwise called network protection, refers to the act of safeguarding our advanced gadgets, information, and organizations from unapproved access, use, burglary, and harm. This incorporates safeguarding our own data, for example, our names, addresses, telephone numbers, government-backed retirement numbers, Visa numbers, and passwords. Advanced security likewise includes safeguarding our gadgets, like our PCs, smartphones, tablets, and switches, from malware, infections, and toxic programming that can damage or capture them.

History of Network Safety

Safety can be followed back to the beginning when the principal PCs were created during the 1940s and 1950s. Around then, PC frameworks were not associated with any outside organizations, and the essential concern was shielding the actual equipment from harm and robbery.

During the 1960s, the US government started to foster PC frameworks for military purposes. These frameworks were intended to store and handle ordered data, and thus, network safety turned into a basic concern. The main network safety measures were created to shield these frameworks from unapproved access and included elements, for example, client verification, access control, and encryption.

During the 1970s, PC networks started to arise, and primary email frameworks were created. With the development of organizations, the gamble of digital dangers expanded, and new network safety measures were grown, like firewalls, interruption identification frameworks, and infection scanners.

The 1980s saw the ascent of PCs, and with it, the primary PC infections. The principal infection, known as the Elk Cloner, was made in 1981 by a secondary school understudy named Richard Skrenta. The infection contaminated Macintosh II PCs and spread through tainted floppy circles.

The 1990s saw the rise of the web, and with it, another period of digital dangers. Programmers started to take advantage of weaknesses in working frameworks and applications to acquire unapproved admittance to PC frameworks. The primary prominent digital assault happened in 1988 when the Morris Worm tainted a great many PCs around the world.

During the 2000s, digital dangers turned out to be more complex and designated. Cybercriminals started to utilize social designing methods, for example, phishing, to fool clients into uncovering delicate data. The ascent of cell phones likewise presented new security challenges, as they turned into a well-known focus for digital attacks.

Why Do We Need Advanced Network Security?

Advanced security is significant in light of multiple factors. To start with, our own data is important to us and other people who might need to involve it for false or criminal purposes. For instance, on the off chance that a programmer accesses our debit card/credit card number and other individual data, they can utilize it to make unapproved buys or commit wholesale fraud. Second, our gadgets and organizations are helpless against assaults that can make us think twice about protection, security, and, surprisingly, our well-being. For example, a programmer can utilize malware to assume command over our gadget or utilize our organization to send off assaults on different gadgets or organizations. At long last, computerized security is vital to safeguard our standing and trust. Having spilt data or hacked gadgets can harm our credibility and associations with others.

In 2020, the Coronavirus pandemic set out new open doors for cybercriminals, as numerous organizations and people moved to remote work and online correspondence. Digital attacks on medical care associations and antibody research offices expanded, and phishing tricks connected with the pandemic became normal.

One more developing worry in network safety is the utilization of computerized reasoning (man-made intelligence) and AI (ML) by the two aggressors and protectors. Man-made intelligence and ML can be utilized to distinguish and answer dangers more rapidly and precisely than people. However, they can likewise be utilized to foster more modern assaults and dodge identification.

The ascent of digital currency (cryptocurrency) has also set out new open doors for cybercriminals. Cryptographic money, also known as cryptocurrency, is frequently utilized in ransomware assaults, as it permits hackers to get payment secretly and without the requirement for a customary ledger. To address these and other arising dangers, network safety is turning into an undeniably significant field of study and business. Legislatures and organizations are putting resources into network safety instruction and preparation, and there is a developing interest in talented online protection experts.

Target: The Digital Strike

Meet Radha, a young professional living in the vibrant city of Bangalore. Like many others, her life revolved around the digital world. From the moment she woke up to the time she lay her head to rest, she navigated a world of social media, online shopping, and digital communication. Little did Radha know that her seemingly ordinary digital activities made her a prime target for hackers. They saw her as a gateway to a gold mine of valuable information. With their cunning skills and insidious techniques, they exploited the vulnerabilities that lay hidden in her digital life.

It began innocently enough. Radha received an enticing email, seemingly from her favourite online retailer, offering an exclusive discount. Unknown to her, this was the first step in a carefully crafted phishing scheme. As she eagerly clicked on the link, a wave of malicious software infiltrated her device, silently collecting her personal data.

Radha's social media profiles were another goldmine for the hackers. They scoured her public posts and photos, gaining insights into her personal life, relationships, and preferences. Armed with this information, they constructed personalized spear-phishing attacks, crafting messages that copied the language and tone of her closest friends.

Through the manipulation of emotions and trust, the hackers tricked Radha into revealing more personal information, such as her passwords and financial details. Unknown to her, the invisible walls of her digital life had been breached, and her once private life is now exposed to those with malicious intent.

This small glimpse into Radha's life highlights the grim reality of how hackers exploit our digital existence. With each click, post, and transaction, we unknowingly offer them the pieces of information they need to invade our lives. It serves as a reminder that in this interconnected world, we must remain vigilant, armed with knowledge, and fortified by cybersecurity measures, to protect ourselves from the invisible threats that sneak into the shadows of our digital lives.

Chapter 2

Cyber-Attack: How Hackers Exploit our Digital Life

Cybercriminals are continually advancing their strategies to take advantage of weaknesses in our hi-tech lives, making significant harm to people, organizations, and even nations. Understanding how programmers exploit these weaknesses is vital in safeguarding our computerized resources and individual data. As technology advances, so do the strategies and methods utilized by hackers to take advantage of our advanced lives. Understanding the strategies utilized by cybercriminals is fundamental to minimizing the dangers related to digital attacks. People and associations should focus on cyber protection.

Now let's see which types of attacks hackers perform to invade our privacy or steal our important data.

1. **Phishing Attacks:** One of the most common methods used by hackers is phishing attacks. They employ various tactics, such as deceptive emails, text messages, or fake websites, to trick unsuspecting individuals into revealing sensitive information like passwords or credit card details. These attacks often appear legitimate, impersonating reputable organizations or individuals. By exploiting human trust and curiosity, hackers gain unauthorized access to personal accounts, leading to identity theft, financial loss, and compromised privacy.

2. **Malware Infections:** Malware, a broad term for malicious software, is another tool commonly employed by hackers. It includes viruses, worms, ransomware, and spyware. Once installed on a victim's device, malware can create havoc by stealing sensitive data, corrupting files, or gaining unauthorized control over the system. Malware often spreads through infected email attachments, compromised websites, or malicious downloads. Hackers capitalize on security vulnerabilities in operating systems and software to exploit the digital lives of unsuspecting users.
3. **Social Engineering:** Hackers are not limited to exploiting technical vulnerabilities alone; they also prey on human psychology. Social engineering techniques manipulate individuals into divulging confidential information or granting unauthorized access. Hackers might impersonate trusted individuals, like tech support personnel or colleagues, and convince their targets to provide sensitive data or login credentials. By exploiting human emotions like fear, urgency, or trust, cybercriminals manipulate individuals to their advantage.
4. **DDoS Attacks:** Distributed Denial of Service (DDoS) attacks target websites or online services, rendering them inaccessible to legitimate users. Hackers achieve this by overwhelming the target's servers with a flood of incoming traffic from multiple sources. By exploiting vulnerabilities in network infrastructure or harnessing botnets, hackers disrupt services, causing financial losses and reputational damage. DDoS attacks are often used as a smokescreen to divert attention from other malicious activities or as a means of extortion.
5. **Zero-Day Exploits:** Zero-day exploits refer to vulnerabilities in software or operating systems that are unknown to the developer or vendor. Hackers discover these vulnerabilities

before they are patched, giving them a significant advantage. By exploiting zero-day exploits, cybercriminals gain unauthorized access, steal sensitive data, or compromise systems. These attacks are particularly dangerous as there is no immediate defence against them until a patch is released, leaving users vulnerable.

6. **Ransomware Attacks:** Ransomware attacks have gained significant attention in recent years. This type of cyber threat involves encrypting a victim's data and demanding a ransom in exchange for the decryption key. Ransomware is often delivered through phishing emails or compromised websites. The impact of ransomware attacks can be devastating, leading to financial loss, operational disruption, and compromised privacy.

Data Security: A Wake-Up Call: Understanding the Impact of Data Breaches

Data breaches occur when unauthorized individuals gain access to confidential data, leading to significant consequences for individuals, companies, and even governments. To understand the seriousness of the problem, let's explore a notable example of a data breach and its implications.

Example: Equifax data breach

One of the most crucial data breaches in recent history is the Equifax data breach, which occurred in 2017. Equifax, one of the major credit reporting agencies in the United States, suffered a cyber-attack that exposed the personal information of approximately 147 million customers. The breach included private data, such as social security numbers, names, addresses, and in some cases, even driver's licence numbers.

Impact on Individuals

The Equifax data breach had severe implications for the affected people. With access to such private data, hackers can engage in identity theft, fraud, and other malicious activities. Cybercriminals can use stolen information to start fraudulent accounts, apply for loans or credit cards, and even file false tax returns. The victims of the breach faced financial losses, damaged credit scores, and had to spend significant time and effort to restore their identities and secure their accounts.

Impact on Businesses

The Equifax breach not only affected individuals but also had far-reaching effects on businesses. Companies depend on credit reporting agencies like Equifax to make informed decisions regarding credibility and risk assessment. The breach exposed vulnerabilities in Equifax's security practices, eroding trust in the company's ability to protect private information. This loss of trust can have a lasting effect on Equifax's reputation, customer base, and financial stability. Furthermore, the breach highlighted the importance of data protection and cybersecurity measures for companies across industries, leading to greater scrutiny and regulatory changes.

Impact on Government and Society

Data breaches of such magnitude also influence governments and society as a whole. Governments must handle the fallout from breaches, pass regulations, and allocate resources to investigate and mitigate the consequences. The Equifax breach spurred conversations about data privacy, cybersecurity laws, and the need for stricter oversight of organizations handling sensitive information. It also stressed the importance of international cooperation in addressing cyber threats, as data breaches often transcend national borders.

Summary

The Equifax data breach serves as a stark warning of the critical need for robust cybersecurity measures and proactive data protection practices. It shows several key lessons:

- ✓ **Strong Security Infrastructure:** Organizations must invest in comprehensive security infrastructure, including robust firewalls, encryption, intrusion detection systems, and employee training, to protect sensitive data effectively.
- ✓ **Timely Response:** Swift and transparent contact is crucial in the aftermath of a breach. Affected people should be promptly notified and provided with advice on mitigating the risks.
- ✓ **Data Minimization:** Collecting and storing only necessary personal information helps reduce the potential effect of a breach. Implementing strict data retention rules can limit the exposure of private data.
- ✓ **Compliance and Regulatory Standards:** Organizations must stay abreast of changing cybersecurity regulations and industry best practices to ensure compliance and protect against possible breaches.

The Equifax data breach serves as a sad example of the widespread effect of data breaches on people, businesses, and society. It underscores the pressing need for enhanced cybersecurity measures, proactive data protection practices, and robust regulatory frameworks. Organizations and individuals alike must remain vigilant, continuously update their security procedures, and prioritize the protection of sensitive data. By learning from past breaches and taking proactive steps to mitigate risks, we can create a safer digital world for all.

The Digital Ninja: Always Behind You

Once upon a time in a dream city called Mumbai, there was a super-smart hacker named Amit. He had a special talent for finding secrets and sneaking into places he wasn't supposed to be. One day, he decided to test his skills on a famous tech company called ABC Corp.

Amit knew that to break into the company's computer systems, he needed to learn as much as he could about them. So, he came up with a plan. First, he went to the ABC Corp cafeteria and pretended to be just another employee having lunch. But really, he was watching the other employees very closely.

As he watched, he saw many of them typing on their laptops and phones to log into their work accounts. Amit paid attention to what they were typing and tried to remember their passwords. He knew that if he could figure out their passwords, he might be able to get into the company's computer systems.

But Amit didn't stop there. He knew that sometimes people throw away important things without realizing it. So, one night, he sneaked into the area where ABC Corp threw away their garbage. This is called 'dumpster diving'. Among all the trash, he found some incredible things like old notebooks and broken electronic devices.

One of the notebooks had important information written in it, like usernames and passwords. Amit couldn't believe his luck! He now had even more ways to get into ABC Corp's computer systems. With all this valuable information, Amit went back to his secret hideout and started putting the pieces together. He connected the usernames and passwords he had gathered and figured out how the company's computer network was set up.

Using this knowledge, Amit was able to sneak into ABC Corp's computer systems without anyone knowing. He felt like a digital ninja, moving from one part of the system to another without leaving a trace. He found secret files, important company secrets, and even personal information about ABC Corp's customers.

But the company eventually found out what Amit had done. Their cybersecurity team discovered the unusual activity on their systems and started investigating. They realized they had been hacked! They quickly worked to stop Amit's access and make their computer systems stronger. They learnt an important lesson about keeping their information safe and protecting themselves from hackers like Amit. Amit, on the other hand, disappeared into the digital world, leaving behind a big mess for ABC Corp to clean up. The company learnt the hard way that they needed to be extra careful with their information and make sure they had strong security measures in place.

So, this story teaches us that even big companies can be hacked if they don't protect themselves well enough. It's important to always keep our passwords safe, be careful with our personal information, and remember that there are people out there who try to find ways to break into our digital lives. Stay safe in the digital world!

Chapter 3

Footprinting: The First Step

A hacker tries to gather information about their target in many ways. This is the first phase of hacking, also known as 'footprinting'. The locally available information may contain sensitive information about the target. Using this method, the hacker collects information like emails, contacts, and domain name information. With the help of some social engineering, the hacker tries to gain more and more sensitive information.

There are two types of footprinting:

1. Active
2. Passive

Active Footprinting

A hacker getting inside the network (digitally and/or physically) is known as active footprinting. The hacker is directly or indirectly connected to the internal network of the target. Different kinds of information that can be gathered from footprinting are as follows:

The operating system of the target machine, firewall, IP address, network map, security configurations of the target machine, email id and password, server configurations, URLs, and VPN.

Various techniques fall into the category of social engineering. A few of them are:

- **Dumpster Diving:** A hacker trying to find out sensitive information in the garbage or dumps is known as dumpster diving. Sometimes, an attacker may find a piece of paper or some important documents in the garbage. When a hacker penetrates a private network, they try to collect as much information as possible.
- **Eavesdropping:** The attacker tries to record the personal conversation of the target victim with someone who is being held over communication mediums like the telephone.
- **Shoulder Surfing:** In this technique, a hacker tries to catch personal information like email id, password, etc. of the victim by looking over the victim's shoulder while they are entering (typing/writing) their personal details for some work.

Passive Footprinting

- **Social Media:** People tend to release most of their information online. Hackers consider such sensitive information seriously. They may create a fake account to establish authenticity and be added as friends or follow someone's account for grabbing their information.
- **JOB Websites:** Organizations share confidential data on many job websites like monsterindia.com. For example, a company posted on a website: "Job Opening for Lighttpd 2.0 Server Administrator." From this, information can be gathered that an organization uses the Lighttpd web server of version 2.0.
- **Google:** Search engines, such as Google, can perform more powerful searches than one can think and one had gone through. It can be used by hackers and attackers to do something that has been termed 'Google hacking'. Basic search techniques combined with advanced operators can do great damage. Server operators exist like 'inurl:', 'allinurl:', 'filetype:', etc.

Google can be used to uncover many pieces of sensitive information that shouldn't be revealed. A term even exists for the people who blindly post this information on the Internet, they are called 'Google Dorks'.

- **Archive.org:** The archived version refers to the older version of the website, which existed a time before, and many features of the website have been changed. archive.org is a website that collects snapshots of all the websites at regular intervals of time. This site can be used to get some information that does not exist now but existed before on the site.
- **An Organization's Website:** It's the best place to begin for an attacker. If an attacker wants to look for open-source information, which is information freely provided to clients, customers, or the general public, then simply the best option is the organisation's website.
- **Who Is:** This is a website that serves a good purpose for hackers. Through this website, information about the domain name, email id, domain owner, etc., anything can be traced. Basically, this serves as a way for website footprinting.

Cyberspace Toolkit

1. **VPN (Virtual Private Network):** A VPN encrypts your Internet connection, ensuring your online activities remain private and secure. It masks your IP address, making it difficult for hackers to trace your online identity.
2. **Password Manager:** Password managers securely store and generate complex passwords for your various online accounts. They help you create strong, unique passwords and remember them, reducing the risk of password-related vulnerabilities.
3. **Antivirus Software:** Antivirus software protects your devices from malware, viruses, and other malicious software. It scans files and programmes, detects threats, and removes them, keeping your system safe from cyber-attacks.
4. **Two-Factor Authentication (2FA):** 2FA adds an extra layer of security to your accounts by requiring two forms of identification. It typically involves a combination of a password and a unique code sent to your phone or generated by an authentication app.
5. **Firewall:** A firewall acts as a barrier between your computer or network and potential threats from the Internet. It monitors and filters incoming and outgoing network traffic, blocking unauthorized access and protecting against intrusions.
6. **Encryption Tools:** Encryption tools encode your sensitive data, making it unreadable to unauthorized users. This ensures that even if your data is intercepted, it remains secure and protected.
7. **Ad Blocker:** Ad blockers prevent intrusive and potentially harmful advertisements from appearing on websites you visit.

They help safeguard against malicious ads that may contain malware or phishing attempts.

8. **Web Browser Security Extensions:** Security extensions, such as HTTPS Everywhere and Privacy Badger, enhance your web browsing experience by encrypting website connections and blocking tracking scripts, respectively.
9. **File Shredder:** File shredders permanently delete files from your device, ensuring that they cannot be recovered. This is particularly useful when disposing of sensitive information or old storage devices.
10. **Backup and Recovery Tools:** Backup and recovery tools automatically create copies of your important files and data, allowing you to restore them in case of data loss due to hardware failure, theft, or ransomware attacks.

These tools play vital roles in safeguarding your digital identity and protecting you against various cyber threats. Remember to keep them up to date and utilize them in conjunction with safe online practices to maintain a strong cybersecurity posture.

Chapter 4

Creating a Strong Digital Fortress: Protecting Your Accounts

Understanding the Risks Associated with Browsing the Internet

Now, we will see what malpractices are done by hackers and how can we protect ourselves from them. The Internet has transformed the way we access information, connect with others, and engage in various activities. While it offers numerous benefits and opportunities, it also presents certain risks that we must be aware of. This beginner's guide aims to provide an understanding of the potential dangers associated with browsing the Internet and offers tips on how to stay safe and secure.

Malware and Viruses

One of the primary risks when browsing the Internet is encountering malware and viruses. Malware refers to malicious software designed to harm your computer or steal your personal information. Viruses, a type of malware, can replicate and spread from one system to another. They can be acquired by downloading infected files, clicking on suspicious links, or visiting compromised websites. To mitigate these risks, it is essential to have reliable antivirus software installed on your device and be cautious when downloading files or clicking on unfamiliar links.

Phishing Attacks

Phishing attacks are attempts to trick users into divulging sensitive information, such as passwords, credit card details, or personal data.

Attackers often pose as trustworthy entities through emails, messages, or websites that appear legitimate. They create a sense of urgency or use social engineering techniques to manipulate users into revealing their information. To protect yourself from phishing attacks, be sceptical of unsolicited messages, double-check the sender's identity, and avoid clicking on suspicious links. When in doubt, contact the organization directly through official channels to verify the authenticity of the communication.

Identity Theft

Identity theft involves the unauthorized use of someone's personal information for fraudulent purposes. When browsing the Internet, you may inadvertently expose personal data, such as your name, address, social security number, or financial details. Cybercriminals can exploit this information to open fraudulent accounts, make unauthorized purchases, or commit other criminal activities in your name. To reduce the risk of identity theft, be cautious about the information you share online, use strong and unique passwords, enable two-factor authentication whenever possible, and regularly monitor your financial accounts for any suspicious activity.

Online Scams

The Internet is widespread with various online scams, targeting unsuspecting individuals. These scams can come in various forms, such as lottery scams, romance scams, or investment scams. Scammers often use persuasive techniques, false promises, or emotional manipulation to deceive their victims. To protect yourself from online scams, exercise caution when dealing with unknown individuals or offers that seem too good to be true. Research and verify the legitimacy of companies or individuals before sharing personal or financial information or making any payments.

Privacy Concerns

When browsing the Internet, your online activities, preferences, and personal information are often collected and tracked by websites, advertisers, and other third parties. This can raise privacy concerns, as your data may be used for targeted advertising, profiling, or even sold to other organizations without your knowledge or consent. To protect your privacy, consider using a VPN to encrypt your Internet connection, regularly review and adjust your privacy settings on websites and apps, and be mindful of the information you share online.

Cyberbullying and Online Harassment

The anonymity and accessibility of the Internet have unfortunately facilitated instances of cyberbullying and online harassment. People can use various online platforms to target and harass others, leading to emotional distress and even psychological harm. To combat cyberbullying and online harassment, it is essential to be mindful of your online behaviour, treat others with respect, and report any instances of harassment to the appropriate authorities or platform administrators.

Fake News and Misinformation

The Internet has become a breeding ground for fake news and misinformation, which can have far-reaching consequences on individuals and society. Misleading information can spread rapidly, leading to misunderstandings, confusion, and the erosion of trust in reliable sources of information. To combat the spread of fake news and misinformation, it is crucial to critically evaluate the sources of information, cross-check facts with reputable sources, and promote media literacy and critical thinking skills.

Social Engineering Attacks

Social engineering attacks rely on manipulating human psychology to deceive individuals and gain unauthorized access to their accounts or sensitive information. Attackers may pose as trustworthy individuals, such as customer support representatives or colleagues, and exploit human tendencies like trust, curiosity, or a desire to help. To protect yourself from social engineering attacks, be cautious of unsolicited requests for personal information, avoid clicking on unfamiliar links or opening attachments from unknown sources, and verify the identity of individuals before sharing any sensitive information.

While the Internet offers a wealth of opportunities, it is essential to understand the potential risks associated with browsing. By being aware of the dangers of malware and viruses, phishing attacks, identity theft, online scams, privacy concerns, cyberbullying, fake news, and social engineering, you can take proactive steps to protect yourself online. Implementing measures such as using reliable antivirus software, being cautious with sharing personal information, staying informed about current scams, safeguarding your privacy, and developing critical thinking skills can significantly enhance your online safety and security. Remember, maintaining a strong digital presence requires ongoing vigilance and adopting a security-first mindset when browsing the vast and ever-evolving landscape of the Internet.

How Can We Protect Our Accounts or Digital Identity?

The first line of defence against unauthorized access is a strong and unique password. As an expert in cybersecurity, I will guide you through the process of creating robust passwords and implementing best practices to protect your accounts from malicious actors.

The Importance of Strong Passwords

Passwords act as the keys to our digital lives, granting access to email accounts, social media profiles, online banking, and more. Unfortunately, many individuals still use weak passwords, making it easier for hackers to gain unauthorized access. A strong password is the foundation of account security and a crucial component in maintaining your digital fortress.

Password Complexity and Length

When creating a password, complexity and length are essential factors. Aim for a combination of upper and lowercase letters, numbers, and special characters. Avoid using common words, personal information, or easily guessable patterns. The longer and more complex your password is, the more resistant it will be to brute-force attacks, where hackers systematically attempt different combinations to crack your password.

Unique Passwords for Each Account

Using the same password across multiple accounts is a grave mistake. If one account is compromised, all your other accounts become vulnerable. It is crucial to use unique passwords for each online service or platform you access. While it may be challenging to remember multiple passwords, the security benefits far outweigh the inconvenience.

Two-Factor Authentication (2FA)

Enhancing password security with two-factor authentication (2FA) adds an extra layer of protection. With 2FA, you provide a second form of verification, such as a unique code sent to your mobile device, a fingerprint scan, or a security token. Even if an attacker manages to obtain your password, they would still need the additional factor to gain access, significantly reducing the risk of unauthorized entry.

Password Managers

Managing numerous unique and complex passwords can be challenging. Password managers offer a solution by securely storing your passwords in an encrypted vault. These tools generate and remember strong passwords for you, eliminating the need to remember them manually.

Popular password managers include LastPass, Dashlane, and 1Password. Choose a reputable password manager that suits your needs, and ensure it has robust security features.

Regularly Update and Change Passwords

Passwords should not remain static for extended periods. Regularly updating and changing your passwords is essential to maintain account security. Aim to change your passwords every three to six months or immediately if you suspect any account compromise. Additionally, promptly update your passwords in case of a known data breach or when recommended by the service provider.

Be Wary of Phishing Attempts

Even the strongest password is useless if you fall victim to phishing attacks. Phishing emails, messages, or websites attempt to trick you into revealing your login credentials or personal information. Exercise caution and be vigilant when opening emails or clicking on links. Verify the legitimacy of websites and double-check email senders before providing any sensitive information.

Secure Your Devices and Networks

Passwords alone cannot protect your accounts if your devices or networks are compromised. Ensure your devices have up-to-date antivirus software, firewalls, and the latest security patches. Secure your home or office Wi-Fi network with a strong password and encryption protocols like WPA2 or WPA3. Avoid using public Wi-Fi networks for sensitive transactions, as they may lack proper security measures.

Regularly Monitor Your Accounts

Maintaining strong passwords and following security protocols is an ongoing process. Regularly monitor your accounts for any suspicious activities, such as unrecognized logins or changes to personal information.

Many online platforms provide security settings and notifications that allow you to monitor and track account activity. Enable these features and review account statements, transaction histories, and notifications to quickly detect any unauthorized access or suspicious behaviour.

Education and Awareness

Cybersecurity threats are constantly evolving, making it essential to stay informed and educated about the latest security practices. Follow trusted sources for cybersecurity news and updates, attend webinars or workshops, and engage in ongoing learning. By understanding the tactics employed by hackers, you can better protect yourself and your accounts.

Regular Backups

In the event of a security breach or ransomware attack, having recent backups of your important data can be a lifesaver. Regularly back up your critical files, documents, and photos to an external hard drive, cloud storage, or a combination of both. This ensures that even if your accounts are compromised, you can restore your data without paying ransom or losing valuable information.

Continuous Security Updates

Operating systems, applications, and plugins require regular security updates to patch vulnerabilities. Enable automatic updates on your devices to ensure you receive the latest security patches promptly. Regularly check for updates manually if automatic updates are not available. Staying up-to-date with the latest security measures is crucial in protecting your accounts from emerging threats.

Building a robust digital fortress requires a proactive approach to password security and account protection. By following the best practices outlined above, such as creating strong and unique passwords, implementing two-factor authentication, and using password managers, you significantly reduce the risk of unauthorized access to your accounts. Additionally, staying vigilant, educating yourself about the latest threats, and regularly monitoring your accounts will help you detect and respond to any suspicious activity promptly. Remember, protecting your accounts is not a one-time effort but an ongoing commitment to safeguarding your digital life. Take the necessary steps today to build a strong defence against cyber threats and protect your valuable information.

Betrayal: The Bank Heist

Once upon a time, in a small town called Manali, there lived a mischievous hacker named Shubham. He was known for his cunning ways and clever tricks. One day, Shubham set his sights on a bank and came up with a devious plan to steal money.

Deepali, an innocent bank employee, was going about her day, helping customers and managing the bank's operations. Shubham knew that if he wanted to get his hands on the bank's money, he needed to manipulate Deepali.

Using a technique called 'social engineering', Shubham began his plot. He pretended to be a bank official and called Deepali, claiming there was a problem with the bank's computer system. He spoke in a calm and confident voice, gaining Deepali's trust.

Shubham told Deepali that he needed her help to fix the problem. He explained that he needed access to the bank's vault to resolve the issue quickly. Deepali, thinking she was assisting a genuine bank official, agreed to help.

As soon as Deepali opened the vault, Shubham seized the opportunity. He swiftly stole three lakh rupees, carefully avoiding any surveillance cameras. Deepali, unaware of Shubham's true intentions, continued to believe she was assisting a bank official.

With the stolen money in his possession, Shubham disappeared into the shadows, leaving behind a confused and unsuspecting Deepali. It wasn't until later that Deepali realized she had been tricked by a cunning hacker.

The bank soon discovered the missing money and launched an investigation. They realized that Shubham had used the power of persuasion and deception to manipulate Deepali and gain unauthorized access to the vault.

The incident served as a lesson for Deepali and the bank. They learnt the importance of being cautious and verifying the identity of anyone asking for sensitive information or access to restricted areas. They also strengthened their security measures, ensuring that social engineering attacks would be more challenging to execute in the future.

As for Shubham, his clever plan worked, but he underestimated the power of trust and the consequences of his actions. Eventually, his unlawful deeds caught up with him, and he faced the consequences of his crimes.

This story teaches us to be careful and sceptical of strangers even if they claim to be someone they are not. Always remember to verify the identity of people asking for personal information or access to restricted areas. By staying alert and cautious, we can protect ourselves and prevent clever hackers, like Shubham, from taking advantage of us.

Chapter 5

Social Engineering: Secret Weapon of Hackers

Social engineering is a strategy used by hackers or attackers to influence and fool people into revealing private information or performing acts that may compromise their security. It involves manipulating human psychology, trust, and social connections to deceive individuals into surrendering access to private data, passwords, or financial details. Social engineering attacks can occur through different means, such as phone calls, emails, instant chats, or in-person conversations. The ultimate purpose of social engineering is to acquire unauthorized access to personal or private information, conduct fraud, or carry out other malevolent activities. An attacker manipulates the person/user to gain sensitive information using social engineering. Social engineering may happen in humans or with the use of tools. Both kinds of attacks are essential, respectively. If a hacker can control the customer services or receptionist of a corporation, they can gain some sort of critical information from there.

Hackers can compromise and get much sensitive information to perform further attacks. A human is the weakest aspect of any corporation. Exploiting the human through manipulation can give a lot of critical information and in some situations gives access to the network of the company. There is no solution to correct human manipulation. Hence, an individual is always vulnerable to social engineering, and the whole business network is vulnerable. Simply influencing an individual can result in massive information disclosure. The person may be directly or indirectly associated with the corporation; maybe a peon or clerk or maybe a manager or an officer at a higher level.

How Exactly Does Social Engineering Work?

There is no such role of the position in social engineering. It fully depends on the manipulating skills of the attacker. Many times, hackers may appear as the lowest post employee like a peon or claim to be at a higher position like an area manager who came for inspection.

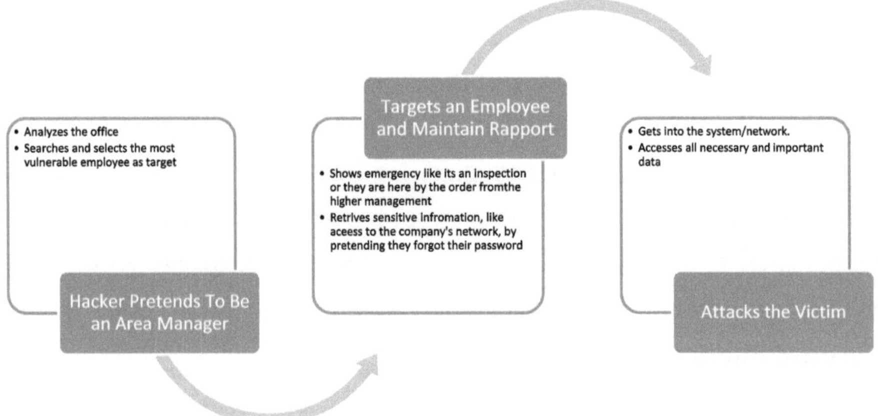

The Procedure of Social Engineering

1. Pretexting

Attackers manipulate trust by creating false scenarios or stories through a social engineering technique known as 'pretexting'. Pretexting involves the creation of a believable and convincing backstory to deceive individuals and gain their trust. Here is how attackers manipulate trust through false scenarios:

a. **Establishing a Plausible Story:** Attackers create a fictional narrative that aligns with the target's environment or situation.

They may pose as a trusted individual, such as an IT support technician or a co-worker, and craft a story that seems legitimate. For example, they might claim to be investigating a security breach or conducting a routine check-up on systems.

b. **Exploiting Human Nature:** Attackers leverage psychological tactics to exploit human tendencies and emotions. They may play on the target's desire to be helpful or have a curious nature or fear of consequences. By appealing to these emotions, attackers create a sense of urgency or necessity, making the target more likely to comply with their requests.

c. **Building Rapport and Trust:** Attackers invest time and effort in building rapport with the target. They may engage in friendly conversation, show empathy, or display knowledge about the target's personal or professional life. This establishes a sense of trust and makes the target more susceptible to following the attacker's instructions or sharing sensitive information.

d. **Providing False Credentials or Documentation:** To further enhance the credibility of their story, attackers may present false credentials or documentation. This could include fake identification badges, official-looking documents, or professional-sounding email addresses. By presenting these artificial facts, attackers aim to reinforce their legitimacy and increase the chances of the target falling for their scam.

e. **Manipulating Communication Channels:** Attackers carefully choose the communication channel to maximize the effectiveness of their deception. They may use phone calls, emails, instant messages, or even in-person interactions to deliver their false scenarios. By leveraging a channel that the target perceives as trustworthy or official, attackers increase the chances of their message being accepted without suspicion.

2. Tailgating and Impersonation

Two prominent techniques in this domain are tailgating and impersonation. This article delves into the world of tailgating and impersonation, exploring their definitions, real-life examples, and measures to prevent falling victim to these manipulative tactics.

Tailgating: Tailgating, also known as piggybacking, refers to the act of an unauthorized individual following closely behind an authorized person to gain physical access to a restricted area. Examples include an attacker entering an office building by closely trailing an employee through a secure access point or slipping into a restricted area during a busy period without presenting proper identification.

Example:

Amit, an employee, is entering the office building, which requires the employees to use their access cards to enter the building and access specific areas, after a lunch break. As he approaches the secured entrance, he notices a person behind him rushing towards the door. Without thinking twice, Amit holds the door open for the person and assumes they are an authorized employee. But it was actually a tailgater, a malicious individual who wants to gain unauthorized access to the building. The tailgater exploits Amit's goodwill and trust, taking advantage of the opportunity to bypass security measures.

Once inside, the tailgater blends in with the crowd, making it difficult for security personnel to distinguish them from legitimate employees. They may proceed to roam freely throughout the building, potentially accessing restricted areas, sensitive information, or even carrying out malicious activities. In this example, the tailgater successfully gained entry to the building without proper authentication. This scenario highlights the risks associated with tailgating, as it allows unauthorized individuals to exploit the trust and assumptions of employees, putting the organization's security at risk.

To prevent such incidents, employees should be vigilant and understand the importance of not allowing unauthorized individuals to tailgate their entry. Strict adherence to access control policies, such as not holding doors open for unknown individuals, can help mitigate the risk of tailgating. Additionally, employees should be encouraged to report any suspicious behaviour to security personnel to ensure the safety and security of the organization.

Impersonation: Impersonation involves attackers posing as authorized personnel to deceive individuals and gain their trust. This can range from impersonating maintenance staff, IT technicians, or even high-level executives. By adopting a guise of legitimacy, attackers exploit the trust placed in these roles to extract sensitive information or gain unauthorized access.

Example:

In a corporate office, Radha, an employee, receives an email that appears to be from the IT department. The email states that there is an urgent security update that requires immediate action. The sender claims to be a trusted IT technician named Vinod and provides a phone number for any inquiries.

Poor Radha, the email is actually a well-crafted impersonation attempt by a cybercriminal. The attacker has carefully researched the organization's employees and identified common names within the IT department to increase the credibility of the impersonation. Trusting the email's legitimacy and feeling a sense of urgency, Radha contacts the provided phone number and speaks to the impersonator posing as Vinod. The impersonator explains that due to the critical security update, they need Radha's login credentials to perform the necessary actions remotely. They reassure her that it is a standard procedure and emphasize the urgency. Feeling pressured and believing she is speaking to a genuine IT technician, Radha unknowingly provides her login credentials to the impersonator. With this information, the attacker gains unauthorized access to Radha's account, compromising sensitive company data and potentially causing further harm.

In this example, the attacker successfully impersonated an IT technician and exploited Radha's trust in the organization's IT department. By leveraging a sense of urgency, knowledge of internal processes, and a well-crafted impersonation, the attacker tricked Radha into disclosing her login credentials.

To prevent falling victim to impersonation attacks, employees should exercise caution and follow best practices, such as:

- ✓ **Verifying Identities:** Always confirm the authenticity of requests by contacting the person or department directly through known and trusted channels.
- ✓ **Double-Checking Email Addresses:** Pay attention to email addresses and look for any discrepancies or suspicious variations.
- ✓ **Being Cautious with Sensitive Information:** Never share login credentials, financial information, or other sensitive data via email or over the phone unless you have confirmed the legitimacy of the request.
- ✓ **Reporting Suspicious Incidents:** If something feels off or raises suspicions, report it to the appropriate authorities within the organization, such as IT or security personnel.

Phishing Attacks: Unmasking the Deceptive Threats Lurking in Your Inbox

Phishing attacks have become one of the most prevalent and damaging cyber threats in the digital environment. These malicious tactics prey on human vulnerabilities by tricking individuals into divulging sensitive information or falling victim to harmful actions. In this article, we will explore the insidious nature of phishing attacks, examining the deceptive emails, fake login pages, and malicious attachments used by attackers to gain unauthorized access and compromise security. Phishing attacks have become increasingly sophisticated and pose a significant threat to individuals and organizations alike. By understanding the methods employed by attackers, such as sending deceptive emails or messages, creating fake login pages, and using malicious attachments, individuals can better recognize and avoid falling victim to these fraudulent schemes.

Sending Deceptive Emails or Messages

Phishing attacks often begin with the delivery of deceptive emails or messages designed to appear legitimate and trustworthy. These emails may masquerade as communications from reputable organizations, colleagues, or service providers. Here is how attackers employ this tactic:

a. **Impersonation:** Attackers impersonate well-known brands, such as banks or online retailers, to create a sense of urgency or credibility. They use official logos, email templates, and even personalization techniques to convince recipients that the communication is authentic.

b. **Social Engineering Techniques:** Phishers leverage psychological manipulation by inducing fear, curiosity, or excitement to entice recipients into taking action. Urgent requests to update account information, claims of unauthorized access, or promises of exclusive offers are common strategies to lure victims.

   Once a victim takes the bait and clicks on a deceptive link or opens an attachment, attackers capitalize on the ensuing vulnerability to gain unauthorized access or compromise security.

Impact of Social Engineering Attacks on Corporate Offices

Social engineering attacks pose a significant threat to corporate offices, exploiting human vulnerabilities to gain unauthorized access to sensitive information. These attacks can have many consequences, affecting both the trustworthiness and reputation of organizations. This article explores the profound impact of social engineering attacks on corporate offices, focusing on data breaches and compromised networks, as well as the financial and reputational damage that ensues.

A. Data Breaches and Compromised Networks

Social engineering attacks can result in devastating data breaches and compromise the integrity of corporate networks. Here are some key points to consider:

Unauthorized Access to Sensitive Data: Successful social engineering attacks grant attackers access to confidential information, including customer data, financial records, intellectual property, and trade secrets. This compromised data can be exploited for malicious purposes, leading to severe financial and legal consequences.

Potential Regulatory Non-Compliance: Data breaches resulting from social engineering attacks can put organizations at risk of violating data protection regulations, such as the General Data Protection Regulation (GDPR) or the California Consumer Privacy Act (CCPA).

Loss of Customer Trust: Data breaches erode customer trust in the organization's ability to protect their personal information. The exposure of sensitive data can lead to identity theft, fraud, or other forms of abuse, negatively impacting customer loyalty and brand reputation.

B. Financial and Reputational Damage

Social engineering attacks can inflict significant financial and reputational harm on corporate offices. Here are some key considerations:

Financial Losses: Organizations incur direct financial losses due to social engineering attacks. These may include expenses related to incident response, investigation, legal actions, regulatory penalties, and potential compensation to affected individuals. Moreover, stolen funds, fraudulent transactions, and unauthorized access to financial accounts can lead to substantial monetary damages.

Reputational Damage: Social engineering attacks tarnish an organization's reputation, as they highlight vulnerabilities and suggest a lack of robust security measures. The negative publicity surrounding a breach can deter potential customers, partners, and investors from engaging with the organization. Rebuilding trust and restoring a damaged reputation can be a long and arduous process.

Operational Disruption: Successful social engineering attacks can disrupt normal business operations, leading to downtime, reduced productivity, and increased recovery costs. For example, if attackers gain control of critical systems or encrypt data through ransomware, organizations may face operational paralysis until the issue is resolved.

Summary

Social engineering attacks have significant and wide-ranging consequences for corporate offices. Data breaches and compromised networks expose organizations to the theft of sensitive information, regulatory non-compliance, and the loss of customer trust. The financial impact includes direct monetary losses, regulatory penalties, and potential legal liabilities. Reputational damage is equally detrimental, as it affects customer loyalty, partnerships, and the overall perception of the organization.

To mitigate the impact of social engineering attacks, organizations must invest in comprehensive cybersecurity measures. This includes implementing robust security protocols, conducting regular employee training and awareness programmes, adopting multi-factor authentication, conducting vulnerability assessments, and establishing incident response plans. By prioritizing cybersecurity and building a culture of vigilance, organizations can fortify their defences against social engineering attacks and protect their valuable assets and reputation.

The Deceptive Call: The Tale of a Fake Amazon Employee

In a small town, there lived a kind-hearted 60-year-old man named Mr. Smith. He loved online shopping and enjoyed using his Mastercard to buy things he needed. Little did he know that a cunning scammer was plotting to deceive him.

One day, as Mr. Smith was sitting at home, his phone rang. A person claiming to be from Amazon introduced themselves as an employee and told Mr. Smith that there was an issue with his account. The person sounded friendly and professional, gaining Mr. Smith's trust.

The fake Amazon employee told Mr. Smith that to fix the problem, they needed some personal information, including his Mastercard number. Now, you would think, "Why would Mr. Smith give his card number to a stranger?"

Well, the scammer was very clever. They used a technique called voice phishing to trick Mr. Smith. By pretending to be from a trusted company like Amazon, they manipulated him into believing that they needed his card number to resolve the issue.

Trusting the person on the other end of the line, Mr. Smith innocently provided his Mastercard number. Little did he know that the scammer had evil intentions.

Days went by, and Mr. Smith continued his daily routine, unaware of the danger looming over him. Then, when he received his credit card statement, he discovered something shocking. A whopping $40,000 had been charged to his card!

Mr. Smith felt a knot in his stomach as he realized he had been deceived. He immediately contacted his bank to report the fraud. The bank investigated the matter and discovered that the fake Amazon employee had used Mr. Smith's card information to make unauthorized purchases.

The scammer had vanished into thin air, leaving Mr. Smith in a state of shock and despair. The money he had worked hard for was gone, all because of the deceitful actions of the scammer.

This story serves as a cautionary tale. It reminds us to always be sceptical of unexpected phone calls, especially from people asking for personal information. We must never share our sensitive details, like credit card numbers, with strangers over the phone.

Remember, if you receive a suspicious call, it's always best to hang up and contact the company directly using their official phone number. Stay vigilant, protect your personal information, and never let the tricks of scammers bring harm to you or your loved ones.

Chapter 6

Phishing, Smishing, and Other Scams: How to Recognize and Avoid Social Engineering Attacks

Phishing, Smishing, and Other Scams

Phishing has grown as one of the most frequent and successful strategies adopted by hackers to fool and abuse people. In the next section, we will dive into the world of phishing, analyzing its deceptive nature, the techniques deployed by attackers, and strategies to protect oneself against this ever-present danger.

Phishing attacks are a sort of social engineering that targets human vulnerabilities rather than attacking technical vulnerabilities. Attackers pose as trustworthy organizations, such as banks, social media platforms, or well-known corporations, to fool users into giving critical information or completing particular activities. The purpose is to get login passwords, financial details, or other important personal or corporate data.

Here are some common techniques employed by phishers:

i. **Fake Login Pages:** Attackers create sophisticated replicas of legitimate login pages, such as those used by banks or popular online services. These pages are designed to trick users into entering their login credentials, which are then captured by the attackers. Unbeknownst to the victim, their information is compromised, leading to unauthorized access to their accounts.

ii. **Malicious Attachments:** Phishers often include malicious attachments, such as infected documents or executable files, in their deceptive emails. When unsuspecting victims open these attachments, malware is deployed onto their devices. This malware can capture keystrokes, steal sensitive information, or even grant remote control access to attackers.

iii. **Spear Phishing:** Spear phishing is a more targeted form of phishing attack that focuses on specific individuals or organizations. Attackers gather personal information about their targets from various sources, such as social media profiles or public databases, to craft customized and believable emails. These messages may appear to come from trusted contacts, colleagues, or business partners. Spear phishing attacks aim to deceive recipients into revealing sensitive information, clicking on malicious links or opening infected attachments.

iv. **Smishing:** Smishing, or SMS phishing, is a type of phishing attack that targets individuals through text messages on their mobile devices. Attackers use social engineering techniques to create urgency or curiosity, luring recipients into clicking on links or providing personal information via SMS replies. Smishing attacks often impersonate well-known organizations, such as banks or government agencies, and may request verification codes, account details, or login credentials.

v. **Vishing:** Vishing, short for voice phishing, involves attackers using voice calls to deceive individuals. They typically pose as representatives from trusted organizations, such as banks or tech support, and employ persuasive techniques to trick victims into revealing sensitive information over the phone. Vishing attacks often leverage social engineering tactics, such as creating a sense of urgency or fear, to manipulate victims into providing personal or financial details.

Preventing Phishing Attacks

Mitigating the risks associated with phishing attacks requires proactive measures and user awareness. Here are a few best practices to protect against these deceptive threats:

- ✓ **Education and Awareness:** Conduct regular security awareness training programmes to educate employees about the tell-tale signs of phishing attacks. Teach them how to identify suspicious emails, URLs, or requests for sensitive information.
- ✓ **Verify Before Clicking:** Encourage individuals to verify the legitimacy of emails and messages by independently contacting the organization or person through trusted contact information, instead of relying on information provided in the suspicious communication.
- ✓ **Implement Email Filters and Security Software:** Utilize robust email filters and anti-phishing software to automatically identify and block phishing attempts before they reach the user's inbox.
- ✓ **Multi-Factor Authentication (MFA):** Enable MFA for all critical accounts and services. This adds an extra layer of protection, making it more difficult for attackers to gain unauthorized access even if credentials are compromised.
- ✓ **Regular Software Updates:** Keep all software, including operating systems and security applications, up to date. This helps protect against known vulnerabilities that attackers may exploit.

Adopting preventive measures, promoting user awareness, and implementing robust security solutions are vital steps in defending against phishing attacks. Stay vigilant and employ best practices.

How Can We Detect a Phishing Page?

There are various techniques to identify a phishing page, such as:

1. Checking the URL of the page and checking for indicators of irregularity, misspelling, or lack of encryption.
2. Comparing the quality of the text and searching for evidence of bad language, spelling problems, or amateurish design.
3. Checking if the content is missing or outdated, such as broken links, expired offers, or absence of contact information.
4. Being suspicious of demands that require personal information, such as passwords, bank data, or social security numbers.
5. Using a fake password or a password manager to test if the website accepts any input or informs you of a mismatch.
6. Checking the payment method and searching for evidence of unusual or unsafe activities, such as wire transfers, gift cards, or cryptocurrency.
7. Using a browser extension or a security programme that can inform you of dangerous pages or block malicious content.

Hacker's Toolkit

Below are some of the tools used by hackers to attack the target.

- **John the Ripper:** John the Ripper is a widely used password-cracking tool that can help hackers uncover weak passwords. It employs various techniques, such as brute-force attacks and dictionary attacks, to crack password hashes.
- **Hashcat:** Hashcat is another powerful tool used for cracking passwords. It supports multiple hashing algorithms and can utilize the power of GPUs (Graphics Processing Units) to accelerate the cracking process. Hashcat is highly flexible and can handle various cracking methods, including brute-force, dictionary, and rule-based attacks.
- **Aircrack-ng:** Aircrack-ng is a popular tool used for Wi-Fi scanning and network analysis. It can capture packets, monitor Wi-Fi networks, and perform various attacks to crack Wi-Fi passwords. Aircrack-ng is often used by ethical hackers and security professionals to assess the security of wireless networks.
- **Kismet:** Kismet is an open-source Wi-Fi scanning and network discovery tool. It passively monitors Wi-Fi networks, captures network traffic, and provides detailed information about nearby access points, connected devices, and network vulnerabilities. Kismet is commonly used for wireless network assessment and troubleshooting purposes.
- **Metasploit Framework:** Metasploit Framework is a powerful and widely used tool by hackers for creating and deploying malware. It provides an extensive collection of

exploit modules, payloads, and auxiliary tools that assist in developing and executing various types of malicious software.

- **Maltego:** Maltego is a versatile tool used by hackers for information gathering and reconnaissance. While it is primarily used for legitimate purposes, such as digital forensics and threat intelligence, it can also be utilized to gather data and create personalized malware attacks by mapping out target networks, identifying vulnerabilities, and crafting tailored payloads.
- **Keyloggers:** Keyloggers are software or hardware tools that record keystrokes on a victim's device. They enable hackers to capture usernames, passwords, and other valuable information entered by the victim, which can be used for malicious purposes.

Chapter 7

The Mobile Menace: Protecting Your Smartphone and Cloud from Cyber-Attacks

Mobile Security

Mobile devices have become an integral part of our lives, connecting us to the digital world and enhancing our productivity. However, with their increasing popularity, mobile cyber-attacks have also surged. From malware and phishing to network spoofing and physical theft, mobile users face a range of threats. In this article, we will explore the landscape of mobile cyber-attacks, the risks they pose, and effective strategies to safeguard your mobile devices and personal information.

Malware and Mobile Viruses

Malware specifically designed for mobile devices has become a widespread concern. Cybercriminals create malicious apps that masquerade as legitimate ones, tricking users into downloading and installing them. These apps can compromise the security of the device, giving attackers unauthorized access to personal data, financial information, or even control over the device itself. To protect against malware, users should only download apps from trusted sources like official app stores, read app reviews and permissions carefully, and consider using mobile security software that can detect and remove malicious apps.

Phishing Attacks

Phishing attacks, which typically targeted email users, have expanded to target mobile users as well. Cybercriminals send fraudulent messages via text messages, social media platforms, or messaging apps, pretending to be trusted entities to deceive users into sharing sensitive information or clicking on malicious links. Due to the smaller screen size and the sense of urgency often associated with mobile devices, users may be more susceptible to falling for phishing scams. To protect against phishing attacks, users should exercise caution with messages and links, avoid clicking on suspicious links, and verify the authenticity of the sender before sharing any sensitive information.

Network Spoofing and Man-in-the-Middle (MitM) Attacks

Mobile devices frequently connect to various networks, including public Wi-Fi networks. This presents an opportunity for cybercriminals to set up fake Wi-Fi networks, known as 'evil twins', to intercept and capture data transmitted by unsuspecting users. MitM attacks allow hackers to eavesdrop on communications, steal login credentials, or even modify the information being transmitted. To protect against network spoofing and MitM attacks, users must be cautious when connecting to public Wi-Fi networks, avoid transmitting sensitive information over unsecured networks, and consider using VPNs to encrypt their data.

Mobile Ransomware

Ransomware, a type of malware that encrypts data and demands a ransom for its release, has also made its way to mobile devices. Mobile ransomware can lock access to the device or encrypt files, making them inaccessible until the ransom is paid. The consequences of falling victim to mobile ransomware can be severe, as mobile devices often store personal photos, documents, and sensitive information. To mitigate the risks of mobile ransomware, users should regularly back up their data, exercise caution when downloading apps or files from untrusted sources, and consider using mobile security software that offers ransomware detection and protection.

Device Theft and Physical Attacks

Physical theft or loss of a mobile device poses a significant security risk. Unauthorized individuals can gain access to personal information, contacts, emails, and stored passwords. Additionally, attackers can exploit vulnerabilities in the device's operating system or use physical methods to bypass security measures like fingerprint or facial recognition. To protect against device theft and physical attacks, users should enable strong authentication methods, such as passcodes, PINs, or biometric authentication, to secure their devices. They should also consider enabling remote tracking and wiping features to locate and protect their devices in case of loss or theft.

Protecting Your Pocket-Sized Powerhouses

As mobile devices become an integral part of our lives, it is crucial to be vigilant about the growing threats of mobile cyber-attacks. By understanding the risks and implementing effective security measures, users can protect their pocket-sized powerhouses and the sensitive information they contain.

Staying safe in the mobile world requires a combination of proactive measures and smart digital habits. Users should always download apps from trusted sources, keep their devices and apps up to date with the latest security patches, and exercise caution when clicking on links or sharing personal information. Additionally, investing in reputable mobile security software can provide an added layer of protection against malware, phishing attacks, and ransomware. These security solutions often offer real-time scanning, malicious app detection, and anti-phishing features to keep users safe from mobile threats.

Furthermore, practising good mobile hygiene is essential. Users should regularly review and adjust their privacy settings, avoid connecting to unsecured Wi-Fi networks, and be cautious when sharing sensitive information through messaging apps or email. In the event of a lost or stolen device, enabling remote tracking and wiping features can help locate the device and protect personal data from falling into the wrong hands. Ultimately, mobile cyber-attacks are a constant and evolving threat. Staying informed about the latest trends and vulnerabilities is crucial to adapt security practices accordingly. Regularly educating oneself about emerging threats and best practices for mobile security can go a long way in preventing attacks and safeguarding personal information.

Remember, mobile devices are not just communication tools or entertainment devices, they are repositories of personal and sensitive data. By prioritizing mobile security and implementing the necessary safeguards, users can enjoy the benefits of mobile technology while minimizing the risks associated with cyber-attacks. Mobile cyber-attacks are a reality of our digital age. As we become increasingly reliant on mobile devices, it is crucial to understand the risks and take proactive steps to protect ourselves. By adopting secure practices, utilizing reputable security software, and staying informed about emerging threats, we can navigate the mobile landscape with confidence and peace of mind.

Cloud Security

The way organizations and individuals store, handle, and access data has been completely transformed by cloud computing. Cloud services are becoming more and more well-liked because of their adaptability, scalability, and affordability. However, along with the benefits come specific security problems that enterprises must overcome to protect sensitive information. Now, let's explore the various safety features provided by cloud computing and discuss techniques to reduce threats and enhance the security of cloud-based environments.

Data Protection and Privacy: One of the key problems in cloud computing is the protection of data and ensuring privacy. When data is stored in the cloud, enterprises forfeit direct control over its physical placement and management. This raises worries about illegal access, data breaches, and compliance with data protection standards. To solve these problems, enterprises should carefully select credible cloud service providers that offer comprehensive security measures, including encryption, access limits, and rigorous data protection rules. Implementing robust encryption mechanisms, both in transit and at rest, can prevent data against illegal access and give an additional layer of protection.

Shared Responsibility Model: Cloud computing runs on a shared responsibility model, where both the cloud service provider and the consumer have distinct security duties. While the cloud provider is responsible for securing the underlying infrastructure, the customer must safeguard their data and applications within the cloud environment. Understanding this distribution of responsibility is vital to provide comprehensive security. Organizations should analyze the security capabilities and compliance certifications of cloud service providers, as well as establish clear communication channels for resolving security concerns and incident response.

Cloud Service and Supply Chain Risks: Cloud computing depends on a complex network of cloud service providers, data centres, and third-party vendors. This interconnection exposes possible risks throughout the supply chain. A security breach or compromise at any point in the chain might have cascading repercussions on the security of the entire cloud system. Organizations should do due diligence when selecting cloud service providers and analyze their security policies, compliance certifications, and contractual duties. Regular audits and monitoring of service-level agreements (SLAs) can assist guarantee that security standards are maintained across the supply chain.

Cloud Application Security: Securing cloud-based applications is critical, as they routinely store and process sensitive data. Organizations should build strict access controls, ensuring that only authorized employees can access and manipulate data within cloud applications. Regularly patching and updating apps, coupled with vulnerability scanning and penetration testing, helps uncover and mitigate potential security flaws. Employing web application firewalls (WAFs) and intrusion detection and prevention systems (IDPS) can give an extra layer of protection against application-level threats.

Compliance and Regulatory Considerations: Cloud computing sometimes entails the storing and processing of data that is subject to industry-specific regulations and regulatory constraints. Organizations must ensure that their cloud infrastructure and service providers comply with applicable standards, such as GDPR, HIPAA, or PCI DSS. Adequate data governance, including data classification, access controls, and audit trails, is necessary to demonstrate compliance and satisfy legal requirements. Collaborating with legal and compliance teams may help negotiate the complicated regulatory landscape and ensure that cloud-based operations fulfil industry-specific security standards.

Summary

As organizations increasingly adopt cloud computing, understanding and tackling its unique security challenges must be kept at the top. By prioritizing data protection and privacy, clarifying responsibilities under the shared responsibility model, managing supply chain risks, securing cloud applications, and ensuring compliance with relevant regulations, organizations can harness the power of the cloud while safeguarding their sensitive information.

The Fake Wi-Fi Trap: A Tale of Online Trouble

A young girl named Vanita loved using her phone to chat with her friends and share pictures on social media. One day, Vanita went to a park with her family and decided to connect to the public Wi-Fi network available there.

Unknown to Vanita, there was a hacker who wanted to cause trouble. The hacker knew how to use a sneaky technique called the 'evil twin'. They created a fake Wi-Fi network that looked just like the real one Vanita wanted to connect to. Vanita innocently connected to the fake Wi-Fi network, thinking it was safe. However, this allowed the hacker to gain access to her phone and personal information. They started misusing Vanita's social media accounts and began posting mean and hurtful things about her. Vanita was shocked and upset when she saw the mean messages and comments online. She didn't understand why someone would want to be so mean and try to hurt her feelings.

Realizing something was wrong, Vanita told her parents about cyberbullying and social media hacking. They immediately contacted the police for help.

The police worked hard to catch the hacker and put an end to his harmful actions. Vanita learnt an important lesson about using secure Wi-Fi networks and being cautious when connecting to public Wi-Fi.

Remember, always be careful when using public Wi-Fi networks. If someone is mean to you or tries to hack into your accounts, tell a trusted adult, like your parents or a teacher. Together, we can make sure the Internet is a safe and enjoyable place for everyone.

Chapter 8

Public Wi-Fi and Other Network Risks: Safeguarding Your Internet Connection

In our interconnected society, access to the Internet has become essential, and public Wi-Fi networks developed as a simple method to stay connected on the go. However, these networks frequently create substantial security risks, since thieves might exploit flaws to intercept important information or conduct attacks. Now, we analyze the potential risks connected with public Wi-Fi and other network environments, describe common attack techniques used by hackers, and suggest practical strategies to preserve your Internet connection and protect your personal data.

Understanding Public Wi-Fi Risks

Public Wi-Fi networks, such as those found in cafes, airports, hotels, and other public places, are useful for consumers seeking an Internet connection. However, they are intrinsically insecure, as they are often unencrypted and lack robust security mechanisms. Cybercriminals can readily intercept data transferred over these networks, gaining access to usernames, passwords, financial information, and other sensitive data. To mitigate the hazards connected with public Wi-Fi, users should be aware of the following:

- ✓ **Man-in-the-Middle Attacks:** Cybercriminals can position themselves between a user and the network, intercepting and potentially manipulating the communication. This allows them to steal login credentials, implant malicious code, or reroute users to bogus websites.

- ✓ **Evil Twin Networks:** Attackers build up false Wi-Fi networks with legitimate-sounding names to trick users into connecting with them. By doing so, fraudsters can intercept data transmitted across the network and get illegal access to personal information.
- ✓ **Rogue Hotspots:** Cybercriminals can also create rogue hotspots that impersonate legal networks. Unsuspecting users connect to these networks, unintentionally exposing their data to attackers.

Secure Connection Strategies

To protect your Internet connection and sensitive data while utilizing public Wi-Fi networks, consider applying the following strategies:

- ✓ **Use Virtual Private Networks (VPNs):** VPNs encrypt your Internet data, establishing a secure tunnel between your device and the VPN server. This ensures that your data remains private and inaccessible to possible eavesdroppers on the public Wi-Fi network.
- ✓ **Verify Network Authenticity:** Confirm the legitimacy of the Wi-Fi network with the establishment or venue offering the connection. Look for official signage or ask staff members for the exact network name. Avoid joining networks with generic names like 'Free Wi-Fi' or 'Public Wi-Fi' as they are more likely to be rogue networks.
- ✓ **Disable Auto-Connect:** Turn off the auto-connect option on your device to avoid automatic connections to open or previously accessed Wi-Fi networks. This offers you control over which networks you connect to, lowering the chance of connecting to rogue networks.

- ✓ **Enable Firewall and Antivirus Protection:** Ensure that your device's firewall is enabled and that you have up-to-date antivirus software installed. These procedures give an additional layer of security against potential dangers.
- ✓ **Encrypt Your Data:** Whenever possible, utilize websites and services that employ encryption (HTTPS) to protect your data during transmission. Look for the padlock symbol in the address bar of your browser to verify a secure connection.
- ✓ **Be Cautious with Sensitive Activities:** Avoid doing sensitive activities, such as online banking or accessing confidential information, while connected to public Wi-Fi networks. Wait until you are on a secure and trusted network before engaging in such activity.

Other Network Risks

Beyond public Wi-Fi networks, more network hazards can undermine your Internet connection and data security. These dangers include:

Home and Small Office Networks: Ensure that your home or small business network is appropriately secured using a strong, unique password for the Wi-Fi network. Regularly upgrade the firmware of your router to address any known vulnerabilities and defend against unauthorized access. Also, try deactivating remote administration functions on your network to minimize potential assaults.

Malicious Websites and Phishing Attacks: Be cautious when accessing websites, particularly ones that ask you to enter personal information or download files. Watch out for symptoms of phishing, such as odd URLs, grammatical problems, and requests for personal data. Maintain up-to-date web browsers and use browser extensions that block dangerous information.

IoT Devices: Internet of Things (IoT) gadgets, such as smart home devices and wearable technology, might add security vulnerabilities to your network. Change default passwords on IoT devices and ensure they are up-to-date with the latest firmware and security fixes. Consider separating IoT devices on a different network to reduce the impact of a potential compromise.

Mobile Network Vulnerabilities: While mobile networks are often more secure than public Wi-Fi, they are not immune to threats. Be cautious when using mobile networks, particularly in congested regions where attackers may set up false cellular networks to capture data. Regularly update your mobile device's firmware and be aware of the permissions granted to apps.

Best Practices for Network Security:

To improve your overall network security, consider applying the following best practices:

i. **Strong Passwords:** Use unique, complicated passwords for all your accounts and networks. Consider using a password manager to securely store and generate passwords.

ii. **Two-Factor Authentication (2FA):** Enable 2FA whenever possible to add an extra layer of security to your accounts. This often includes entering a verification code provided to your mobile device in addition to your password.

iii. **Regular Updates:** Keep your devices, operating systems, programmes, and antivirus software up to date. Updates often include security updates that address vulnerabilities.

iv. **Regular Backups:** Regularly back up your vital data to an external or cloud storage solution. In the event of a security breach or data loss, backups ensure that you can restore your data and limit the harm.

v. **Educate Yourself:** Stay informed on the latest security threats and techniques employed by cybercriminals. Regularly educate yourself on emerging dangers and best practices for network security.

Summary

Safeguarding your Internet connection is important in today's digital environment. Public Wi-Fi networks and other network contexts present unique security threats, but by implementing the tactics discussed in this article, you may protect your personal data and maintain a secure online presence. Whether it's using VPNs on public Wi-Fi, confirming network legitimacy, or setting strong passwords and security measures for your home network, taking proactive efforts to secure your Internet connection will go a long way in preserving your digital life.

The Secret Shop of Scammers: Lily's Shopping Scam

There was a girl named Lily who loved to shop online. She enjoyed browsing through websites and looking at beautiful clothes and accessories. One day, she came across a website that claimed to have amazing deals and discounts. Excited, she decided to make a purchase.

Lily entered her personal information and credit card details on the website. She eagerly waited for her package to arrive, but weeks went by and there was no sign of it. She tried reaching out to the company's customer service, but they never responded. That's when Lily realized she had fallen victim to an online shopping fraud.

Feeling upset and worried, Lily told her parents about what had happened. They immediately contacted their bank and explained the situation. The bank froze the credit card and started investigating the fraudulent charges.

Lily learnt an important lesson about online safety. She realized that not all websites are trustworthy, and it's essential to be cautious when shopping online. Her parents helped her report the fraud to the authorities, and they worked together to get her money back.

From that day forward, Lily became more careful about where she shopped online. She always checked if the website was secure and reputable before making a purchase. She shared her story with her friends and also reminded them to be cautious.

Remember, when shopping online, make sure to only use trusted websites and never share your personal information with unknown or suspicious sites. Stay safe and enjoy your online adventures!

Chapter 9

Online Shopping and Banking: Tips for Secure Transactions

Online shopping and banking have revolutionized the way we make purchases and manage our finances. However, the convenience of these digital platforms also comes with security risks. Cybercriminals constantly seek to exploit vulnerabilities and gain unauthorized access to personal and financial information. In this article, we will explore essential tips and best practices to ensure secure transactions while shopping and banking online. By following these guidelines, you can protect your sensitive data, safeguard your financial accounts, and confidently enjoy the benefits of online transactions.

Shop and Bank on Secure Websites

When engaging in online transactions, always ensure that you are on secure websites. Look for the padlock symbol in the address bar, and check that the URL begins with 'https://'. The 's' indicates that the website has an SSL (Secure Sockets Layer) certificate, which encrypts data transmission between your browser and the website's server. Encryption helps protect your information from interception by cybercriminals.

Keep Your Devices Secure

Maintaining the security of your devices is crucial for safe online shopping and banking. Keep your operating system, web browsers, and antivirus software up to date with the latest security patches. Regularly install updates to fix vulnerabilities that cybercriminals may exploit.

Additionally, use strong, unique passwords for your devices and online accounts, and consider enabling two-factor authentication (2FA) for an added layer of security.

Be Cautious of Phishing Attempts

Phishing is a common technique used by cybercriminals to trick individuals into revealing their personal and financial information. Be wary of unsolicited emails, text messages, or phone calls asking for sensitive data or directing you to click on suspicious links. Legitimate organizations will never ask you to provide sensitive information through unsecured channels. Always verify the authenticity of the communication and contact the institution directly if you have any doubts.

Use Secure Payment Methods

When making online purchases, opt for secure payment methods. Credit cards often offer better fraud protection than debit cards. Additionally, consider using secure online payment systems, such as PayPal or Apple Pay, which provide an extra layer of security by keeping your financial information private during transactions.

Monitor Your Accounts Regularly

Keep a close eye on your bank and credit card statements. Regularly review your transactions to identify any unauthorized or suspicious activity. If you notice anything unusual, report it to your bank or financial institution immediately. Many banks and credit card companies also offer transaction alert services, which can notify you of any activity on your account in real-time, helping you detect and address potential fraudulent charges promptly.

Use a Virtual Private Network (VPN)

When conducting online transactions, especially while using public Wi-Fi networks, consider using a VPN. A VPN encrypts your Internet connection, making it more difficult for cybercriminals to intercept your data. It provides an additional layer of privacy and security, ensuring that your sensitive information remains protected.

Summary

Online shopping and banking offer convenience and accessibility, but they require caution and security measures to protect your personal and financial information. By shopping and banking on secure websites, keeping your devices updated and secure, being cautious of phishing attempts, using secure payment methods, monitoring your accounts regularly, and using a VPN, when necessary, you can enhance the security of your online transactions. Stay vigilant, follow these tips, and enjoy the benefits of online shopping and banking with peace of mind.

CyberShield: Abhimanyu's Battle Against Hackers

There was a smart and curious boy named Abhimanyu. He loved playing with computers and exploring new things about technology. One day, Abhimanyu heard about hackers who could break into computer networks and cause trouble. He wanted to protect his own network, so he came up with a brilliant idea.

Abhimanyu had heard about something called artificial intelligence and machine learning. These were like super-smart computer programmes that could learn and make decisions on their own. He thought, 'Why not use this amazing technology to defend my network from hackers?'

So, Abhimanyu started learning about AI and machine learning. He read books, watched videos, and even asked his computer-savvy friends for help. With lots of practice and hard work, he built his very own AI-powered defence system. He named his creation 'CyberShield'. CyberShield was like a digital superhero that could watch over Abhimanyu's network all the time. It could detect when a hacker was trying to break in and stop them in their tracks. With CyberShield on guard, Abhimanyu felt safe and protected. Whenever a hacker tried to sneak into his network, CyberShield would quickly identify their sneaky tactics and block them. It was like having a trusty sidekick to help him fight off the bad guys!

Thanks to his imagination and determination, Abhimanyu became a hero in his own right. He showed the world that even a young boy could use the power of artificial intelligence and machine learning to keep his network safe from hackers.

And so, Abhimanyu continued his journey of learning and exploring, always staying one step ahead of the cyber villains, making the digital world a safer place for everyone.

Chapter 10

The Future of Digital Security: Emerging Technologies and Trends to Watch

Digital security is now the most important thing in a world that is becoming more interconnected. To remain ahead of hackers and protect sensitive information, it is necessary to pay attention to developing technologies and trends that hold the key to the future of digital security. Now, we will go into some of the most promising advancements in the industry and highlight the need for proactive actions in securing our digital ecosystem.

I. Artificial Intelligence (AI) and Machine Learning (ML) in Security

Artificial intelligence and machine learning technologies have already made important contributions to different industries, and their potential in digital security is no exception. Here are a few ways AI and ML are shaping the future of digital security:

i. **Threat Detection and Response**

   AI-powered algorithms can analyze vast amounts of data to identify patterns and anomalies, enabling proactive threat detection. Machine learning models can continuously learn from new data to enhance accuracy and efficiency in real-time, identifying and mitigating security threats.

ii. **Behavioural Biometrics**

By analysing user behaviour patterns, AI algorithms can create unique user profiles and detect anomalies that may indicate unauthorized access attempts. Behavioural biometrics, such as keystroke dynamics and mouse movements, offer an additional layer of authentication and enhance overall security.

iii. **Automated Security Operations**

AI and ML technologies enable the automation of security operations, including threat monitoring, incident response, and vulnerability management. By reducing manual efforts and response times, organizations can proactively address security incidents and free up resources for higher-value tasks.

II. Zero Trust Architecture

Traditional security models often relied on perimeter-based defences, assuming that internal networks were secure. However, with the rise of cloud computing, remote work, and interconnected systems, a new approach known as 'Zero Trust Architecture' has gained traction. Key aspects of this approach include:

i. **Identity and Access Management**

Zero Trust Architecture emphasizes granular access controls, strong authentication mechanisms, and continuous monitoring of user activity. Users are only granted access to the resources they require, limiting potential attack vectors and reducing the impact of compromised credentials.

ii. **Microsegmentation**

Networks are divided into smaller, isolated segments, limiting lateral movement within the system. This segmentation helps contain potential breaches, as attackers find navigating through a network with restricted access to specific resources more challenging.

III. Internet of Things (IoT) Security

Here are some critical factors for IoT security:

i. **Device Authentication and Encryption**

Robust authentication mechanisms, including secure protocols and cryptographic techniques, should be implemented to ensure the integrity and confidentiality of IoT communications. Encryption of data at rest and in transit is essential to protect sensitive information.

ii. **Firmware and Software Updates**

IoT devices should have the capability to receive regular firmware and software updates to address security vulnerabilities. Manufacturers must prioritize security updates and provide mechanisms to streamline the update process for end users.

iii. **Network Segmentation and Monitoring**

Isolating IoT devices within dedicated network segments helps contain potential compromises. Continuous monitoring of IoT devices for unusual behaviour and network traffic analysis can aid in the early detection of security incidents.

IV. Privacy and Data Protection

The exponential growth of technology and interconnected systems has raised concerns about how personal data is collected, stored, and used by various entities. Now, we see the significance of privacy and data protection, the challenges they pose, and the importance of safeguarding individuals' personal information in a rapidly evolving digital environment.

Here are a few areas of focus:

Privacy by Design

Building privacy considerations into the design and development of systems and applications ensures that data protection measures are in place from the outset. Privacy-enhancing technologies, such as differential privacy and secure multiparty computation, help preserve privacy while enabling valuable data analysis.

i. **Data Minimization and Retention**

   Collecting and retaining only necessary data reduces the risk associated with data breaches. Organizations should implement data minimization and retention policies that outline the appropriate duration for retaining data and ensure its secure disposal when no longer needed.

ii. **Privacy Regulations and Compliance**

   With the introduction of regulations like the General Data Protection Regulation (GDPR) and the California Consumer Privacy Act (CCPA), organizations must prioritize compliance with privacy requirements. Implementing robust data protection measures, conducting regular privacy impact assessments, and providing transparency to individuals regarding data collection and usage are crucial components of compliance efforts.

iii. **User Education and Consent**

   Empowering individuals with knowledge about privacy best practices and their rights regarding their personal data is essential.

User consent should be obtained explicitly and transparently, and individuals should have the ability to control how their data is used.

Summary

The future of digital security lies in the continuous adaptation and integration of emerging technologies and proactive security measures. AI and ML bring advanced threat detection capabilities, while Zero Trust Architecture ensures a more resilient and secure environment. Securing the IoT and prioritizing privacy and data protection are crucial aspects of the evolving digital landscape.

Here Begins Our Cyber Security Journey

As the cybersecurity environment continues to evolve, organizations and individuals must stay vigilant, keeping abreast of emerging technologies, industry best practices, and regulatory requirements. By embracing these trends and investing in robust security measures, we can navigate the complexities of the digital world with greater confidence, safeguarding our valuable information and ensuring a secure future for all.

Here are my contact details:

Gmail: yogeshukey1723@gmail.com
LinkedIn: Yogesh Ukey
Instagram: @cybersec.yogesh